DINOSAUR
COUNTING BOOK
1 to 20

NOTE: There is a guide to pronouncing the
dinosaur names at the back of this book.

DINOSAUR

COUNTING

BOOK
1 to 20

Frances Mackay

1

ONE sleepy **PTERANODON**.
How would you wake him up?

2

TWO friendly **APATOSAURUS** dinosaurs.
How many tails can you count?

3

THREE smiling **TRICERATOPS** dinosaurs. How many horns does each one have on their head?

4

FOUR hungry **MONTANOCERATOPS** dinosaurs. How many legs does each one have?

5

FIVE baby dinosaurs.
How many have **orange** on their body?

6

SIX dancing dinosaurs.
How many heads can you count?

7

SEVEN fierce **TYRANNOSAURUS REX**.
How many small ones are there?

8

EIGHT scared dinosaurs.
Who do you think will escape
from the T-Rex?

9 NINE funny dinosaur shadows. Which shadow matches this dinosaur?

10

TEN running **COELOPHYSIS** dinosaurs.
How many tails can you count?

11

ELEVEN strange **PARASAUROLOPHUS** dinosaurs. How many are facing RIGHT? →

12

TWELVE mini dinosaurs.
How many are standing on two legs?

13

A swimming ELASMOSAURUS with THIRTEEN spots on his back. How many spots are on his two fins?

14

A cheeky **STEGOSAURUS** with **FOURTEEN** plates on her back. How many spikes are on her tail?

15

FIFTEEN cute dinosaurs hatched today.
How many are hiding behind the bush?

16

A happy mother dinosaur with **SIXTEEN** babies. How many are facing LEFT? ←

17

A gigantic **DIPLODOCUS** with **SEVENTEEN** stripes on her body. Count them!

18

A spiky **KENTROSAURUS** with **EIGHTEEN** spots. How many spots are green?

19

NINETEEN strong **ANKYLOSAURUS** dinosaurs. How many more are needed to make 20?

20

TWENTY colourful dinosaurs.
Which one do you like best?

Now you can count from 1 to 20.

1 2 3 4 5 6 7

8 9 10 11 12

13 14 15 16

17 18 19 20

Well done!

About the dinosaurs in this book

The dinosaurs in this book are, of course, cartoon ones. Here is some information about the real dinosaurs.

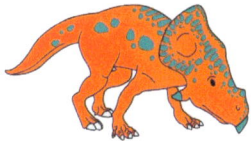

Cartoon dino	What it looked like	Information
		PTERANODON (tuh-ran-uh-don) The Pteranodon wasn't actually a dinosaur. The Pterodactyl and Pteranodon were flying reptiles called Pterosaurs. Wingspan up to 7m (23ft)
		APATOSAURUS (a-pato-sawr-us) Scientists used to think the Apatosaurus and the Brontosaurus were the same, but now they think the Apatosaurus was larger with a thicker neck. Length 23m (75ft)
		TRICERATOPS (tri-ser-a-tops) This dinosaur had three horns - two on its head and one on its snout. It had beak-like jaws and a neck frill. It was about the size of an elephant. Length 9m (30ft)
		MONTANOCERATOPS (mon-tan-o-sair-a-tops) A small, plant-eating dinosaur. It had teeth in its upper jaw. Length 2.5m (8.2ft)
		TYRANNOSAURUS REX (tie-ran-o-sore-us-rex) A fierce dinosaur, commonly known as T-Rex. It had a huge head and a long, heavy tail. It had forward-pointing eyes. Length 13m (42ft)

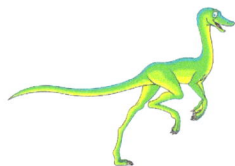

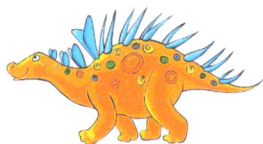

Cartoon dino	What it looked like	Information
		COELOPHYSIS (see-lo-fy-sis) A small, slender meat-eating dinosaur. It had a long neck and tail and was a fast runner. Length 3m (9.8ft)
		PARASAUROLOPHUS (par-ah-saw-rol-o-fus) This dinosaur had a strange-shaped head with a large curved crest. It walked on two and four legs. Length 9.5m (32ft)
		ELASMOSAURUS (ee-las-mo-sawr-us) This wasn't actually a dinosaur but a Plesiosaur. It lived in water. It had four paddles for swimming and a very long neck. Length 10m (34ft)
		STEGOSAURUS (steg-o-sawr-us) This dinosaur had kite-shaped plates along its back and spikes on its tail. The front legs were much shorter than the back ones. Length 6.5m (21ft)
		DIPLODOCUS (di-plod-ocus) A gigantic plant-eating dinosaur with a long neck and a long, whip-like tail. Length 26m (85ft)
		KENTROSAURUS (ken-tro-sawr-us) A small Stegosaur with plates and spikes along its back. It walked on all fours. Length 4m (13ft)

Cartoon dino	What it looked like	Information
		**ANKYLOSAURUS** (an-key-low-sawr-us) A large dinosaur with huge plates of body armour and a large tail club that was strong enough to break bones. Length 8m (26ft)
		**SPINOSAURUS** (spine-o-sawr-us) This was a huge dinosaur with a sail-like structure on its back. It had long jaws with sharp, pointed teeth. Length 18m (59ft)
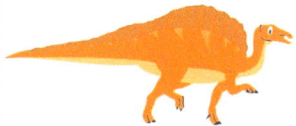		**VELOCIRAPTOR** (vel-o-si-rap-tor) This dinosaur had feathers and was about the size of a turkey. It had a long tail and a sharp claw on each hind foot. Length 2m (6.6ft)
		**POLACANTHUS** (pol-a-can-thus) A plant-eating armoured dinosaur with many spikes on its back and tail. It grew to 5m (16ft) in length.
		**OURANOSAURUS** (or-an-o-sawr-us) A plant-eating dinosaur with a sail-like structure on its back. Length 8.3m (27ft)
		**IGUANODON** (i-gwa-no-don) A large plant-eating dinosaur, as big as an elephant. It had a large thumb spike and a long tongue. Length 10m (33ft)

ABOUT THE AUTHOR

Frances Mackay is the author of more than 90 teacher resource books and has written several picture books, activity books and information books.

She was a primary school teacher for 20 years in Australia and the UK. She loves writing books that make learning fun!

facebook:
@francesmackaychildrensauthor
instagram:
@frances.mackay_author

Read more about Frances and grab some FREEBIES at: www.francesmackay.com

Frances now lives in Tasmania, Australia.

DID YOU ENJOY THIS BOOK?

Your feedback helps me provide the best quality books and helps other readers like you discover great books.

You can leave a review online or send it direct to me at:
frances@francesmackay.com

I read and appreciate each one.

THANK You! ☺

Dinosaur Colouring Pack
www.francesmackay.com

Grab your FREE
10-page
Dinosaur Colouring Pack
when you visit
www.francesmackay.com

BOOKS BY FRANCES MACKAY

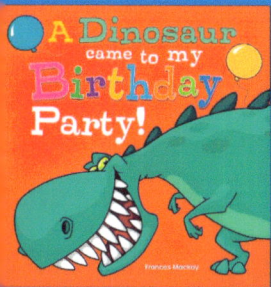

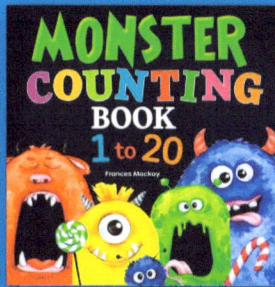

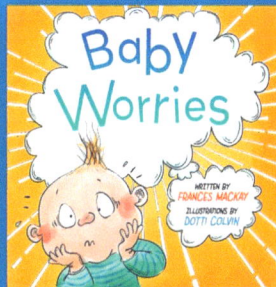

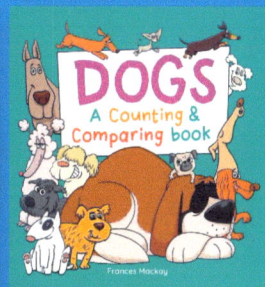

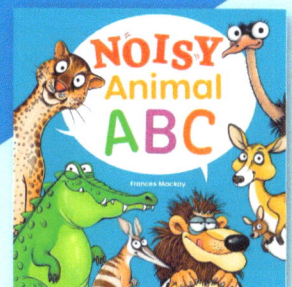

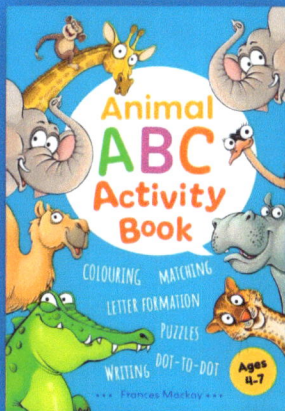

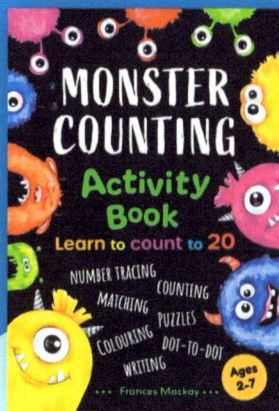

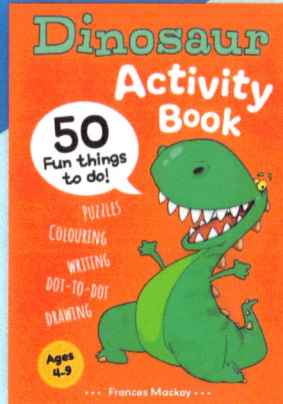

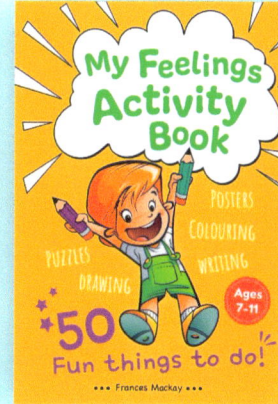

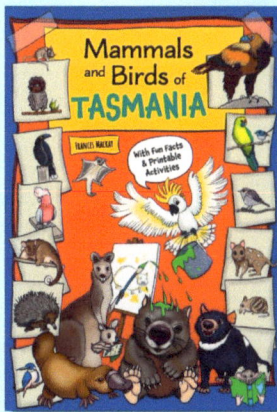

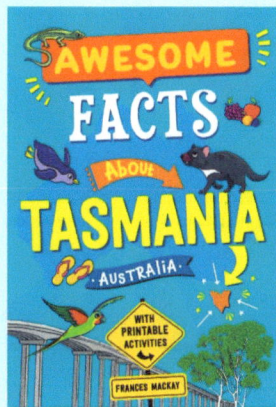

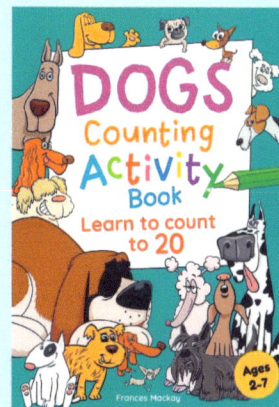

A Dinosaur came to my Birthday Party!
Frances Mackay

MONSTER COUNTING BOOK 1 to 20
Frances Mackay

Baby Worries
WRITTEN BY FRANCES MACKAY
ILLUSTRATIONS BY DOTTI COLVIN

DOGS A Counting & Comparing book
Frances Mackay

NOISY Animal ABC
Frances Mackay

Animal ABC Activity Book
COLOURING   MATCHING
LETTER FORMATION
PUZZLES
WRITING   DOT-TO-DOT
Ages 4-7
Frances Mackay

MONSTER COUNTING Activity Book
Learn to count to 20
NUMBER TRACING
MATCHING   COUNTING
COLOURING   PUZZLES
WRITING   DOT-TO-DOT
Ages 2-7
Frances Mackay

Dinosaur Activity Book
50 Fun things to do!
PUZZLES
COLOURING
WRITING
DOT-TO-DOT
DRAWING
Ages 4-9
Frances Mackay

My Feelings Activity Book
POSTERS
PUZZLES   COLOURING
DRAWING   WRITING
Ages 7-11
50 Fun things to do!
Frances Mackay

Mammals and Birds of TASMANIA
FRANCES MACKAY
With Fun Facts & Printable Activities

AWESOME FACTS About TASMANIA AUSTRALIA
WITH PRINTABLE ACTIVITIES
FRANCES MACKAY

DOGS Counting Activity Book
Learn to count to 20
Ages 2-7
Frances Mackay

Illustration credits
Most illustrations are licensed from Dreamstime - www.dreamstime.com

Other illustrations:
Montanoceratops
https://commons.wikimedia.org/wiki/File:Montanoceratops_BW.jpg
Nobu Tamura (http://spinops.blogspot.com), CC BY 3.0 <https://creativecommons.org/licenses/by/3.0>,
via Wikimedia Commons

Diplodocus
https://commons.wikimedia.org/wiki/File:Diplodocus_carnegii.jpg
Fred Wierum, CC BY-SA 4.0 <https://creativecommons.org/licenses/by-sa/4.0>, via Wikimedia Commons

Velociraptor
https://commons.wikimedia.org/wiki/File:Fred_Wierum_Velociraptor.png
Fred Wierum, CC BY-SA 4.0 <https://creativecommons.org/licenses/by-sa/4.0>, via Wikimedia Commons

www.ingramcontent.com/pod-product-compliance
Lightning Source LLC
Chambersburg PA
CBHW041553040426
42447CB00002B/172